Interpreting Compression Transducer Waveforms – (Including Comp-Pe

Interpreting Compression Transducer Waveforms

User's Manual for the (Comp-Peek-Xducer-Probe)

by Mandy Concepcion,
Automotive Diagnostics and Publishing

www.autodiagnosticsandpublishing.com

Copyright © 2014, 2015 By Mandy Concepcion, Automotive Diagnostic and Publishing

This book is copyrighted under Federal Law to prevent the unauthorized use or copying of its contents. Under copyright law, no part of this work can be reproduced, copied or transmitted in any way or form without the written permission of its author, Mandy Concepcion.

The Comp-Peek-Probe information, schematics diagrams, documentation, and other material in this book are provided "as is", without warranty of any kind. No warranty can be made to the testing procedures contained in this book for accuracy or completeness. In no event shall the publisher or author be liable for direct, indirect, incidental, or consequential damages in connection with, or arising out of the performance or other use of the information or materials contained in this book. The acceptance of this manual is conditional on the acceptance of this disclaimer.

About ASE Certification

We at Automotive Diagnostics and Publishing promote the ASE Certification program and encourage all beginning and advanced technicians alike to participate and get certified. We are not directly affiliated with ASE. ASE Automotive Technician Certifications are recognized throughout the United States by almost all county and state agencies as meeting the requirements to be considered an automotive technician. Many organizations and repair facilities nationwide have ASE Certification as mandatory for their technicians as part of their business model.

It's one thing to just show up for work; it is quite another to take control of your career, and get recognition for the knowledge and skills you've developed over the years. That's what ASE certification is all about: helping you tell customers, employers, and other people about what you know. After all, being an automotive technician is not just turning wrenches. Your years of hard work show that you've gone the extra mile, and put in the time and effort to learn your trade. Your ASE certification patch is proof of it. Get certified now."

We are not directly affiliated with ASE. The Tech-2 is a registered trademark of Vetronix Corp. and GM The DRB III & Starscan are a registered trademark of DaimlerChrysler The NGS is a registered trademark of Ford Motor Co. The Comp-Peek-Probe is copyrighted to Mandy Concepcion, Automotive Diagnostics and Publishing. Any other proprietary name used in this book was done purely for explanatory purposes.

Developed in the U.S.A.

Preface & Acknowledgments

This book, Interpreting Compression Transducer Waveforms – (Including Comp-Peek-Probe) of which there are also video based training, is a dedicated work centered in automotive electric diagnostics. About 99% of the auto-repair duties for an auto technician will be of an electronics nature. In this particular case, the pressure transducer has become a major auto diagnostic support technology. By running simple compression waveforms tests, you will know all there is to know about the mechanical integrity of the engine. Now, with the included free oscilloscope software you don't even need to buy expensive equipment. Here, we'll show you about our Comp-Peek-Probe transducer and then, how to interpret the waveforms. The interpretation section can be used with our Comp-Peek-Probe or any other compression transducer.

About our "Signalator" technology: We are the developer of Signalator, a technology involving software algorithms that interpret the signal waveform and output a diagnostic report, charts and diagrams for the issue at hand. The Comp-Peek-Probe was developed precisely for the Signalator algorithms, since we needed a baseline from which the software could be based on. Signalator is part of the Scope-1 System and available by itself as a separate software.

This book and our other electronics courses are a must have for any technician or DIY interested in delving into the world of electric and electronic auto repair.

Special thanks given to all the techs and shops who supported my investigations on this technology, and to my wife for her support…God bless…

Thanks also given to:

Benjamin Komnick and Sean Boyle for their insight work on compression signals at the <u>Southern Illinois University Carbondale.</u>

(For further insight on the operation and testing of these components see some other complimentary books & DVD-Videos at Amazon.com or our website www.autodiagnosticsandpublishing.com.)

Technician@autodiagnosticsandpublishing.com

Table of Contents

Interpreting Compression Transducer Waveforms..3
    About ASE Certification..7
    Preface & Acknowledgments...9
The Waveform 16
    Why we created the Automotive Compression Peek Transducer Probe................16
    Why you need the Compression Transducer..16
    Solution to your problem..16
    The benefits of the Automotive Compression Transducer....................................17
    About the Comp-Peek-Probe-SCOPE kit...17
    What you get with the Comp-Peek-Probe..18
    The Transducer Itself..18
    The Signal Conditioning Box..21
    The Included Oscilloscope Software..25
How to Use the Comp-Peek-Probe..26
    Anatomy of an Engine Compression Waveform..28
    Compression Waveform at idle...29
    Compression at idle with the Comp-Peek Scope Software..................................29
    Explanations about the waveforms at idle..30
    Facts About Analyzing Compression Waveforms..30
    Dividing the Compression Signal Now these division lines also correspond to the 4 cycles on the internal combustion engine. The cycles are: 31
    Peak-compression Wave Spot..34
    Identifying the Waveform's 4 Cycles...35
    Compression Lowermost Point...36
    The Exhaust Section of the Compression Waveform..37
    The Intake Section of the Wave..39
    Some More Insights..40
In-Depth Event-to-Event Compression Waveform Analysis 43
    In-Depth Basic Points...45
    Deep Analysis...48
    The first point of interest is Point-A:..49
    Point of interest Point-B and Point-H:..49
    Issues with Valve Train..51
    Exhaust Level Analysis, Point D..52
    Then Point – D & E, including the Ex section:..53
    Points F thru G with IN included is the intake hump:..58
Included Comp-Peek-Probe Oscilloscope Software...61
    The Comp-Peek-Probe Software..63
    Detailed Scope Software Explanation 64
    Signal Generator Section Output..65
    Amplitude Section..66
    Base and Time Trigger Levels..67

- A Final Footnote about the Vac-Peek-Probe...68
    - Why we created the Automotive Vacuum Transducer Probe (Vac-Peek-Probe).......................68
    - Why you need the Vacuum Transducer..68
    - Solution to your problem..69
    - The benefits of the Vac-Peek Transducer...69
- About Us and Mandy Concepcion...73
    - AUTOMOTIVE DIAGNOSTICS & PUBLISHING...74
    - Who is Automotive Diagnostics & Publishing..74
    - Innovative Introductions..75

The Waveform

I will start this book with one caveat, **it is the relationship of the compression waveform data-points that matters, not so much each of the specific value alone**.

Modern automotive diagnostics and the advancements in electronics has unleashed various automotive technologies in the field of diagnostics. The compression or high pressure probe is one of them. We at Automotive Diagnostics and Publishing created the Vac-Peek and Com-Peek Probe. In this manual, we'll concentrate on the Comp-Peek Probe.

Why we created the Automotive Compression Peek Transducer Probe

The Comp-Peek automotive compression transducer is a dynamic compression sensor developed for the Auto Scope 1, but can also be used with your own Scope. Why for the Auto Scope 1? Well, the Scope 1 has a special software algorithm called "Signalator", which is an automated fault recognition feature. You set the Scope 1 and it'll record, capture, analyze and point the issue to you without any intervention on your part. So, we needed a special set of specification baseline to work with Signalator. Other than that, it'll perform what it was meant to do on your own scope, minus the Signalator automatic fault capture of course.

Why you need the Compression Transducer

Using the good old compression gauge is fine; there's nothing wrong with it. However, a compression gauge gives you a maximum reading, not the running minute compression signature that this compression transducer does. This scope add-on sensor peeks into the engine in a very deep way. With it, you can determine intake and exhaust valve opening, seal, weak springs and virtually anything related to the valve train. How hard is it to read the signal? It take a bit of practice, but the compression peek transducer probe comes with it's own easy to read manual. Within a few days you'll get the hang of it.

Solution to your problem

Ever had that pesky "Misfire" problem that won't go away? That's what the compression peek transducer probe is for. A misfire could be caused by ignition, injection or mechanical. Often times it's better to do an ignition test or an injector leak-down clog test, when accessible. But often it will take much longer to do any of these diagnostic tests. If you only had a compression gauge, a weak or crappy valve seal won't show up on the gauge, but it'll pop-up "Right Away" on the compression peek transducer probe signal output. That is the power of a transducer like this one and at a price that won't break the bank.

The benefits of the Automotive Compression Transducer

The benefits of using the compression peek transducer are quicker and cleaner diagnostics of pesky misfire codes. Determining after another repair shop's tune-up procedure what the true cause of the issue is and verifiable with the signal. This add-on to your own Scope is the difference between properly tracking a misfire or not. It'll pay for itself in the first repair.

About the Comp-Peek-Probe-SCOPE kit

The newly developed Comp-Peek-Probe Oscilloscope hardware, with included free software is an industry first. We know most of you don't have the resources to invest thousands on an Oscilloscope just for when you use it. So, we went ahead and developed digital circuitry just for the Comp-Peek-Transducer-Probe. This oscilloscope circuitry/hardware is only meant to be used with the Comp-Peek-Probe, not by itself. The Comp-Peek-Probe can be used with any oscilloscope without the Scope, but the scope part of it is meant only for the Comp-Peek-Probe. So you can't use the included oscilloscope on the probe with the Scope-Kit by itself.

Here we created the right electronic circuitry to work with the Comp-Peek-Probe. The Comp-Peek-Probe-Scope box and software runs on your Windows XP, 7, 8 laptop/desktop and has the right filtering circuitry and signal conditioning needed by the Comp-Peek-Probe. The included free software gets the Comp-Peek-Probe signal and shows you the waveform, which you can save to the hard-drive if you want. The Comp-Peek-Probe Scope is a dual channel unit that you can synch to the desired cylinder. So, with this package you're set to diagnose and analyze automotive compression signals.

What are the Comp-Peek-Probe-Scope specifications? It is a 16 bit data acquisition software and the rest doesn't matter because it was developed and made for the Comp-Peek-Probe and nothing else. It is a custom made oscilloscope system for our Comp-Peek-Probe.

What you get with the Comp-Peek-Probe

The Comp-Peek-Probe kit is composed of the Comp-Peek-Probe unit, an oscilloscope software (CD ROM), a 14mm spark plug adapter, a signal conditioning box and a BNC to Banana plug adapter in case your scope uses banana plugs. The signal conditioning box is meant to protect your PC in case the impossible happens, like the probes gets zapped by a spark-plug wire high voltage.

The Comp-Peek-Probe is made of aluminum, so it is rugged and tough. However, it is also a delicate piece of equipment. It is not meant to take shocks, so don't drop it and treat it with care. The probe is a piezoelectric unit, so by definition, it is a crystal device (piezo mean crystal construction).

The Transducer Itself

The Comp-Peek-Probe also comes with a signal conditioning box. This box serves a simple task, which is to adjust the signal strength to accommodate almost any oscilloscope out there. The signal conditioning box has two cables coming out of it, with BNC connectors. It also has an adjustment knob at the end of the box. Both cable BNC connectors are labeled "Scope" and "Probe" for your benefit.

The Signal Conditioning Box

To use the signal conditioning box simply adjust the knob until you get an "unclipped" waveform at the top. A clipping will show up as a flat line portion on the waveform. You don't want any "flat clipping" at the top of the wave.

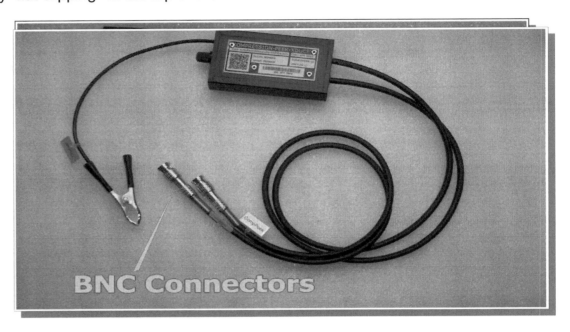

The Comp-Peek-Probe is made tough. It has a solid construction with bolted body and piping. However, this is a sensitive pressure transducer, so don't drop it. It you do it could crack the pressure sensor and you will have to send it back to us for repairs. There is also an extended warranty which you could get for the Comp-Peek-Probe even years after purchase.

Interpreting Compression Transducer Waveforms – (Including Comp-Peek-Prob

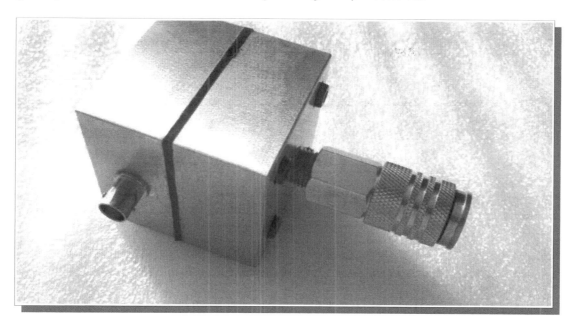

Finally there's a set of hose and spark plug adapters that come with the unit. The spark plug adapter is a 14mm thread, which is the most common spark plug out there. There are many spark plug adapters that can also be purchased on-line though Amazon and others that can save you lots of money, as opposed to getting them through a tool truck. Often you can use spark plug savers with the included hose for a few dollars. So, don't spend too much on spark plug adapters if you don't already have them.

As a reminder, the Comp-Peek-Probe is also Gasoline and Diesel enabled. All you need is to attach the right diesel adapters (not included) to it and you're set.

Interpreting Compression Transducer Waveforms – (Including Comp-Peek-Prob

If you need other sizes like 18mm Ford, then you can purchase an inexpensive 14mm to whatever else you might need. These adapter are found at most tool hoses and online.

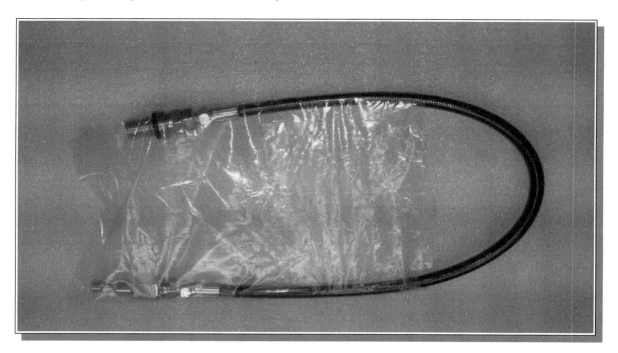

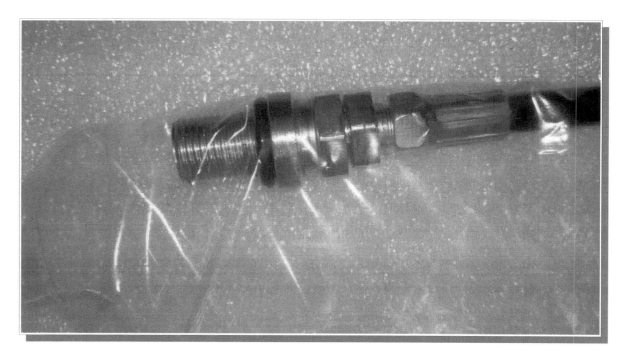

Interpreting Compression Transducer Waveforms – (Including Comp-Peek-Prob

The Included Oscilloscope Software

As you can see, the Comp-Peek-Probe is a complete unit. It is also bundled with an included oscilloscope software. This software is a basic but powerful scope software, which includes synching through channel 2, all normal scope adjustments, and cursors. You also have the ability to save your waveforms to dist for future reference.

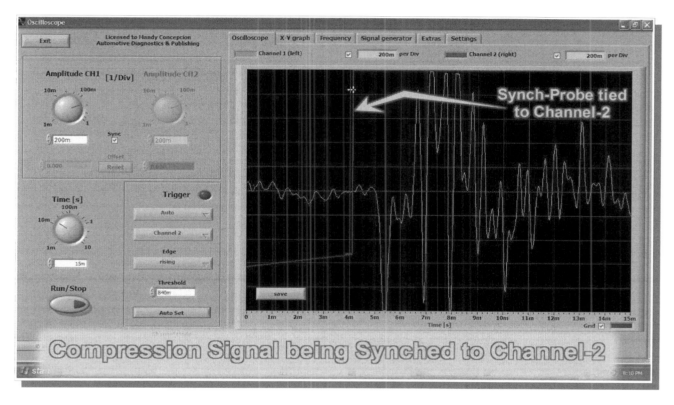

Next, I'll show you how to use the Comp-Peek-Probe from a waveform interpretation point of view. You can also use this knowledge with any other similar pressure transducer. The included oscilloscope (only included with oscilloscope package) software is a great tool built into the Com-Peek-Transducer. It basically uses your sound card as an analog to digital converter to acquire the signal that way you don't have to purchase extra expensive electronics. This is a complete package and you need nothing else to capture the waveform for your Comp-Peek-Probe, other than a PC, desktop or Laptop.

How to Use the Comp-Peek-Probe

Interpreting Compression Transducer Waveforms – (Including Comp-Peek-Prob

Anatomy of an Engine Compression Waveform

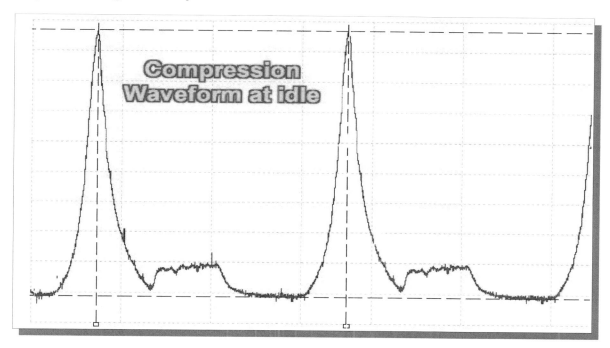

Compression Waveform at idle

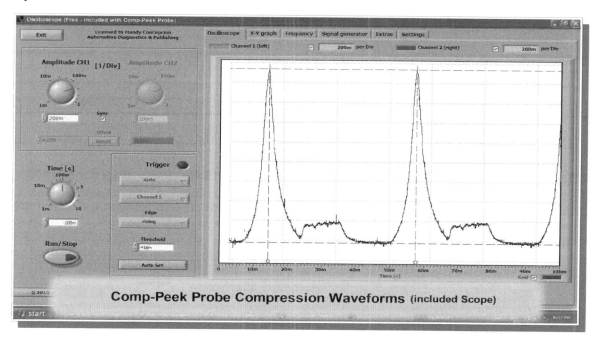

Compression at idle with the Comp-Peek Scope Software

Explanations about the waveforms at idle

Here is a typical compression waveform at idle taken with the Comp-Peek-Probe. You might ask yourself, are these compression signal waveforms the same for all cars? The answer is, yes, as long as they're four cycle engines. And as far as we all know, all automotive engines are 4 cycle engines.

The second figure shows the same waveform using the included Comp-Peek-Probe Oscilloscope (without Signalator). Although not the Scope-1, it is a very capable scope software and comes with the Comp-Peek-Probe-Scope kit. The Comp-Peek-Probe can also be purchased without the Scope circuitry.

With the Comp-Peek-Probe or any other compression transducer the engine can be checked for camshaft to crankshaft timing problems, electronic variable camshaft timing issues, intake and exhaust valve sealing faults, valve spring, piston ring seals, worn cam lobes, restricted exhaust, ignition timing, and cylinder misfire issues. Obviously the list includes most of the hardest diagnostic issues out there. With the Comp-Peek-Probe or using any other compression probe, diagnosing these tough issues will be an automatic habit by just an understanding of the pressure changes that occur within the engine.

We'll begin by analyzing the anatomy of an engine compression waveform at idle. At higher RPMs things get weird due to the high engine speeds and the usability of the signal goes way down. So, without further ado, here's analyzing a compression waveform.

Facts About Analyzing Compression Waveforms

We'll start by dividing the compression signal waveform into four separate sections. The divisions happen at 180 degrees of crankshaft rotation. So, the crankshaft rotates halfway, that's 180 degrees. If it rotates one complete revolution, then it's 360 degrees, basic geometry. Now, the camshaft rotates at half the rate of the crankshaft. So with that in mind, we can ascertain that the crankshaft rotates twice, for every camshaft rotation. That is basic four-cycle internal engine combustion geometry. As I've mentioned, we'll divide the waveform into 180 degree intervals, as seen next.

Interpreting Compression Transducer Waveforms – (Including Comp-Peek-Prob

Dividing the Compression Signal Now these division lines also correspond to the 4 cycles on the internal combustion engine. The cycles are:

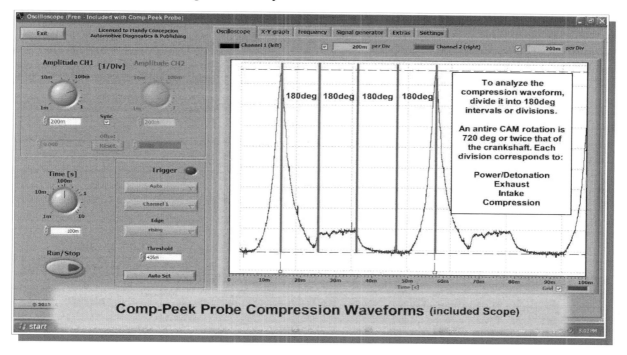

Comp-Peek Probe Compression Waveforms (included Scope)

1) Power/Detonation
2) Exhaust
3) Intake
4) Compression

The first line corresponds to the peak-compression point right before the spark, which causes the mixture to detonate or explode.

Interpreting Compression Transducer Waveforms – (Including Comp-Peek-Prob

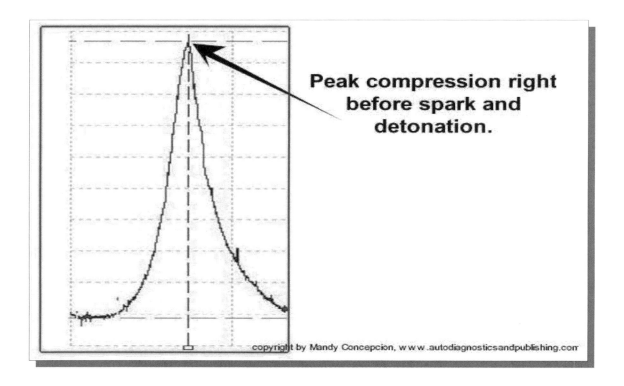

Peak-compression Wave Spot

This is the line you use to determine if you have proper compression. It is not a "How much compression in PSI I have". Some software will output a calculated compression number value, but that's the wrong approach and is misleading. What's important is the ratio or the relationship of this line along the waveform to the other measured points. In other words, how does this key point compares to the other measurable key points, as explained next.

In real life, you can expect a value of around 60 PSI for most engines. That's because there's a huge different between static compression or what you see using a compression gauge, and the dynamic compression using the transducer or Comp-Peek-Probe. The dynamic reading is way more exacting and shows you real life values on the fly. The static way is fine as a generalized idea, but often times misleading by itself. Static compression reading will also not reveal lesser valve leaks and various mechanical issues, due to it's averaging nature. Also the gauge is not too sharp to begin with.

Interpreting Compression Transducer Waveforms – (Including Comp-Peek-Prob

Identifying the Waveform's 4 Cycles

Here we can see the easily identifiable 4 cycles as they correspond to the compression waveform. Remember that all the signal waveform specifics have all to do with the travel of the piston and it's relationship to the valve-train.

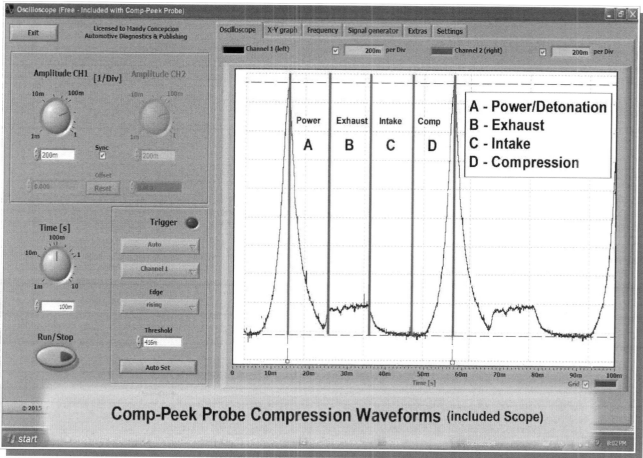

As you can see the next point of interest is the detonation or power down-curve of the waveform. In this case, we're taking the reading with the engine at idle and the particular cylinder disabled, so the spark-plug has been removed and the one ignition coil disabled somehow.

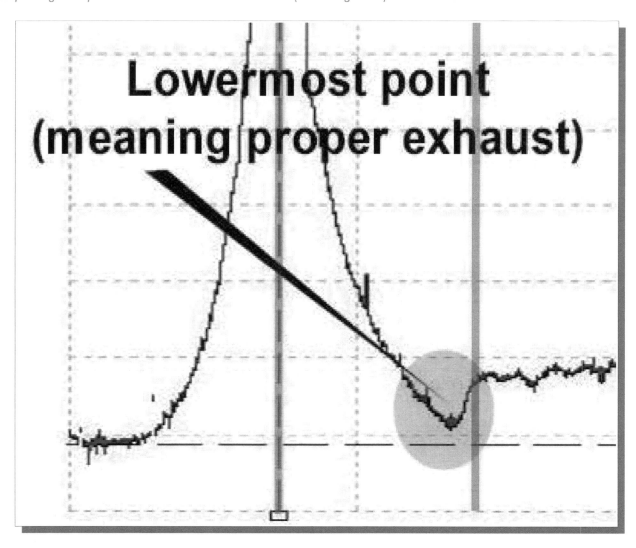

Compression Lowermost Point

So, what we actually see here is a pressure signal curve, not necessarily the mixture exploding. But the compression reading seen here is a mirror image of the working cylinder. A downward curve seen next is expected, and it points to the general health of the exhaust system. If you had a **clogged-catalytic converter**, you'll see it here. The bottom of this curve, right before the second 180 deg. line has to reach it's bottommost point, as seen before the upward going pressure hump started. That low point signifies proper exhaust.

The Exhaust Section of the Compression Waveform

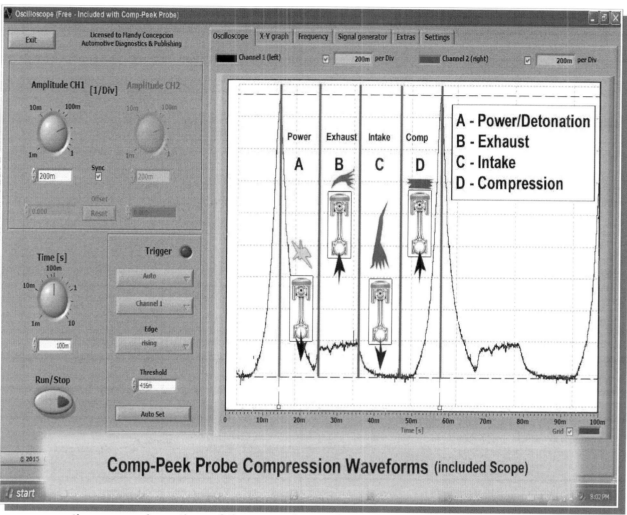

Then comes the second section of the waveform, the exhaust section. It is described as a mesa or flat top hill. Ideally, this section should rise to about 1/6th to 1/9th of the peak compression spike. This is the cylinder in the exhaust cycle. Issues associated with this section of the waveform are clogged catalytic converter and exhaust valves not opening completely usually due to a variable valve timing issue.

Interpreting Compression Transducer Waveforms – (Including Comp-Peek-Prob

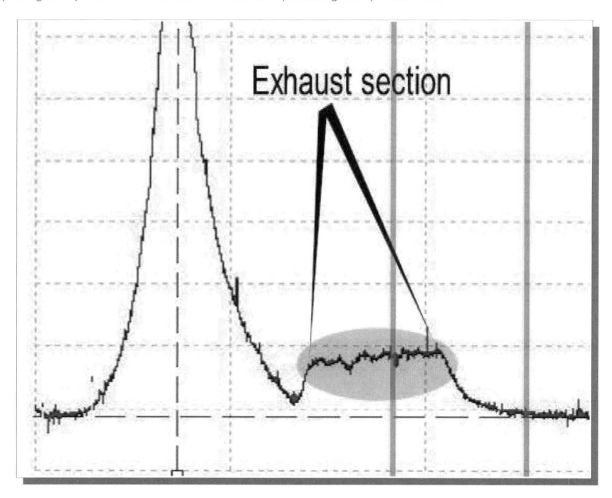

Next, Is the **intake** part of the compression waveform. The intake part of the waveform is denoted by a quick down-slopping followed by a flat line. The down-slopping section points to the remnants of exhaust gases leaving the combustion chamber and then a specific amount of vacuum created due to the downward moving piston. This is also important and will vary a bit depending on whether you goose the throttle or not during the compression test. In general this section will tell you if you have good vacuum created at the cylinder. As a general rule this section should be as low as possible on the scope. This should be the lowermost spot on the wave. It should match the exhaust valve open event, if all is well. Restrictions at the intake throttle body butterfly plate, severely clogged air filter, turbo-waste issues, variable intake valve control system or anything else that will affect the air intake will raise this line.

As I've explained before, **it is the relationship of the compression waveform data-points that matters, not so much each of the specific value alone**. How one section of the waveform looks is worthy of notice, but how all the waveform data points or event relate to each other is king.

The Intake Section of the Wave

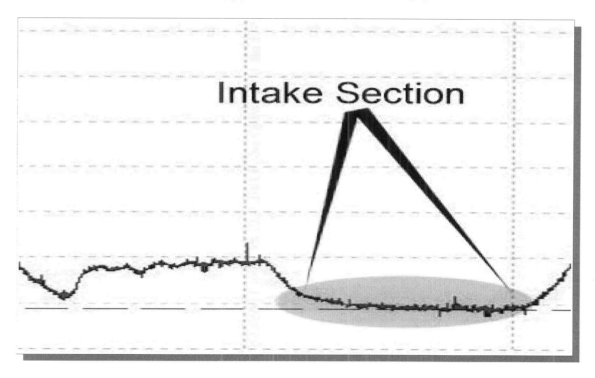

Finally, there's the compression section. This section is categorized by an upward sloping raise of the pressure content in the cylinder. It is then finalized at the uppermost point in the complete waveform and proceeded by the combustion/power cycle.

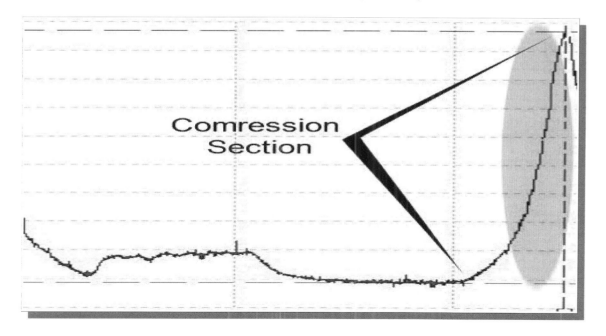

Some More Insights

The compression section has to be at least 8 to 9 times the height of the exhaust signal level. Here's where you know or determine piston and valve train seals, how well the cylinder can contribute and hold pressure and is a precise indicator of engine health. Issues with variable valve timing will also cause the compression section's highest-most spot to be low. Remember that the Comp-Peek-Probe tests are dynamic, meaning as they happen, and for that matter are stressed tests. You're not just doing a static test with a gauge. You're really seeing what the engine is doing mechanically, as it is doing it, on the fly in real time. There's no substitute for this level of insights.

In-Depth Event-to-Event Compression Waveform Analysis

Interpreting Compression Transducer Waveforms – (Including Comp-Peek-Prob

In-Depth Basic Points

A very basic but important factor to comprehend here is that all 4-cycle internal combustion engines have 2 TDC and 2 BDC events for every completed 4 cycle. In other words, every time there's a cylinder detonation or that the spark plug fires, the crankshaft spins twice around and the camshaft once. This fact is often lost when looking at all the ups and downs of the waveform. Above you can see the complete waveform as included in the ICE 4 cycles.

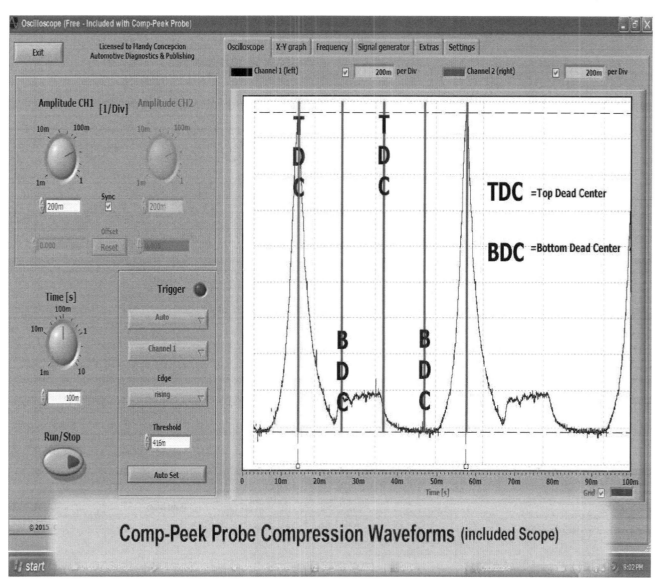

Interpreting Compression Transducer Waveforms – (Including Comp-Peek-Prob

Following, we'll go into a deep analysis of the compression waveform, which may not always be necessary, but may present you with knowledge only found here.

Deep Analysis

The following analysis will show you a series of data-points corresponding to specific events on the compression cycle. This is what "SIGNALATOR" does, it is our algorithm to automate the analysis of the compression waveform. SIGNALATOR follows the compression waveform exactly and analyzes all the data-points to arrive on a diagnostic decision. The Comp-Peek-Probe was created so that the "Signalator" software knows the sensor's baseline and generates repeated results. Each compression probe on the market behaves differently. Signalator needs to base itself on one specification only to diminish erroneous readings. Here's the basis of what we've already learned as a summarized waveform data-point even diagram.

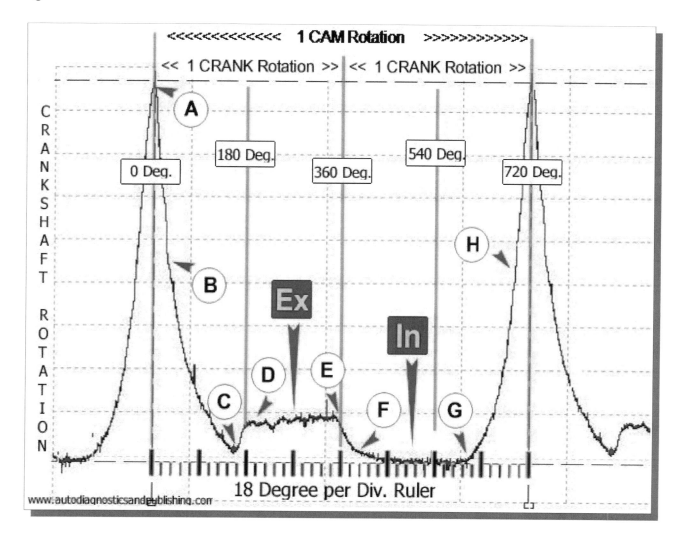

We'll start by explaining the above compression waveform chart. The chart above is divided into the entire 4 cycles of the internal combustion engine, which involves one CAM rotation and two CRANK rotations. Every 360 Deg. The crankshaft rotates and every 720 Deg. the CAM does a complete turn. That is so in every automotive engine.

We can also see the familiar divisions every 180 degrees that we've been using here from the beginning. We also see at the bottom of the chart, a 40 mark ruler that divides the waveform further into 18 Degree increments. So, using the bottom ruler, each small mark equals 18 degrees and every 5 marks equals 90 Deg and every 10 marks equals 180 Deg. Simple geometry, but now we can see where all these compression events fall.

The first point of interest is Point-A:

Point-A is the peak compression point. We've talked about this point before and suffice it to say that Point-A is the beginning of the waveform and should be $1/6^{th}$ to $1/9^{th}$ of Ex (exhaust) or the average exhaust value. Ex is considered the signal between D and E. Remember that modern variable valve timing has a huge influence on compression, so any issues with the compression waveform could also mean that the valve timing electronic control is faulty.

Additionally and also mentioned before, don't concentrate so much on the compression or PSI value of the waveform. Some engines will show 75 PSI during a dynamic test like this, while others 95 PSI and they're both perfect. It is this relationship between the different compression events that matter.

Point of interest Point-B and Point-H:

Point-B is the vertical halfway point of the peak compression hump. This point is important and should be **within 18 to 24 degrees of the TDC** that preceded it. The same goes for Point-H. If point B or H are disparate towards TDC, then there is a mechanical (or variable valve) issue.

Interpreting Compression Transducer Waveforms – (Including Comp-Peek-Prob

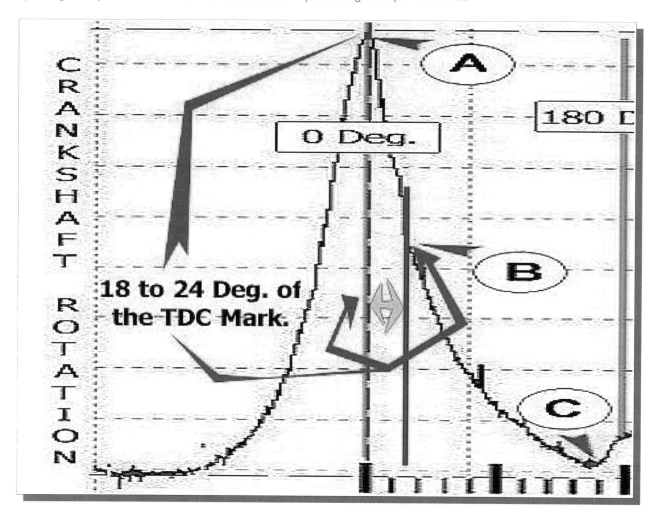

Clearly the **oscilloscope cursor feature** is a plus in this case. Cursors are vertical and also horizontal lines used to perform measurements. With all the hoopla about automotive oscilloscopes having 10MHz or 50MHz, that supposed feature is a complete ripoff for the automotive technician. Modern automotive oscilloscopes won't even need a 1 MHz range to operate very effectively. Why? Because automotive signal operate in at the audio or Kilohertz range. So, having a cursor measurement feature is way more desirable than high frequency range.

Next you can see what an off center compression peak waveform looks like. The issues for this to show up on the waveform are varied, but do point to a mechanical issue.

Interpreting Compression Transducer Waveforms – (Including Comp-Peek-Prob

Issues with Valve Train

And as we've reiterated many times before, issue with Point-B or H make more sense when correlating it to the rest of the waveform, so continue reading. As the piston continues to travel downward, we reach **Point – C:**

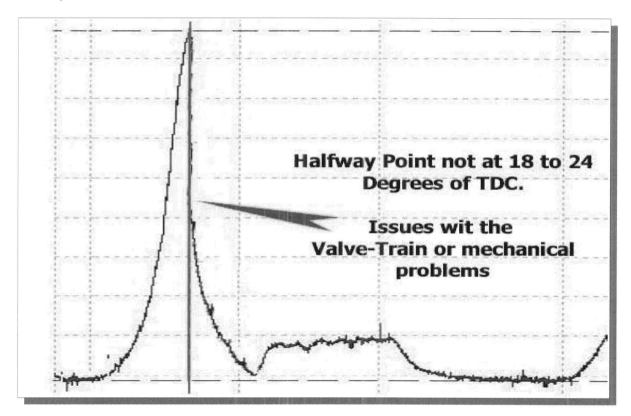

Point – C is the beginning of the exhaust valve opening event. This is a vary important point. The very bottom of this point should be sharp and also should be the lowest in the signal waveform. Also, point C should show up consistently with little variation while the engine is running. This point is right before (split second) the exhaust valve open, so there it vacuum already in the combustion chamber, hence the low value. Point C should also be very close to the **IN point**, which is the **intake cycle** corresponding to engine vacuum. Point C will be higher if there is an issue with intake or exhaust valve seal. Most commonly the intake or exhaust valves have burnt seats or pitted valve rims that won't show up on a compression test using a regular gauge. Point C is precisely why the Comp-Peek-Probe was created or why you should use any form of compression transducer.

Exhaust Level Analysis, Point D

After Point C we then see the combustion chamber returning to the same level as the exhaust pressure. This is the same pressure level as the exhaust system and will be analyzed next. Point C should be around 22 to 48 degrees before BDC. It is a good indicator of CAM timing. Points C and D are also used to denote exhaust CAM timing.

Next, **Point – D** the exhaust valve opens and the pressure balances out to the exhaust system.

There are some worthy facts about Point D worth noting. The first 180 degree BDC mark should fall somewhere between the Points C and D. If the BDC mark is to the left of point C, then exhaust valve timing is retarded. But if the BDC mark falls to the right of point D, then the exhaust valve timing is over advanced. Clearly these issues point out to either a jumped timing belt/chain or a stuck variable valve timing hydraulic pulley or solenoid. Generally speaking, if no the timing belt, then issues with the newer valve timing control.

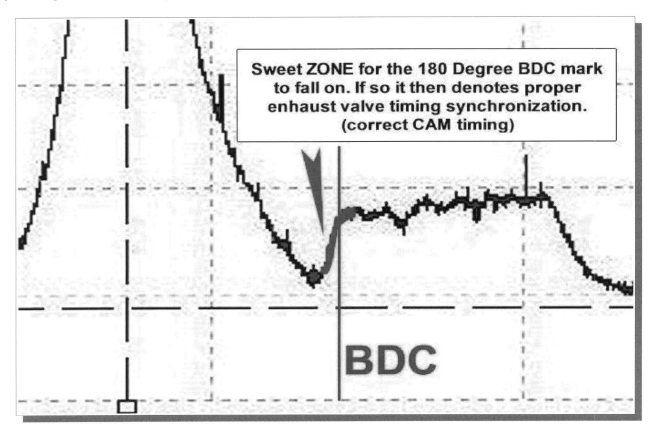

Interpreting Compression Transducer Waveforms – (Including Comp-Peek-Prob

Note that we're also assuming here that we are not dealing with some exotic valve timing cycle, like Miller or Atkinson, which is not normally done at idle. Atkinson cycle, used on Prius and Ford/Nissan hybrid derived engines, uses a delayed valve timing scheme to control compression ratio. In the Miller cycle, the intake valve is left open longer than it would be, therefore skewing the traditional compression transducer reading.

Then Point – D & E, including the Ex section:

The section that includes points D thru E and Ex is the average exhaust pressure cycle. It should be about $1/6^{th}$ to $1/9^{th}$ of the peak compression event, at the beginning of the waveform. It is the same as the pressure at the exhaust system.
The exhaust average value is a great indicator of clogged catalytic converter or general restriction at the exhaust. It can also indicate issues (rare) with the exhaust valves not opening properly, as in worn CAM lobes. But most importantly even are modern electronically controlled valve timing issues. These issues however will skew not just the exhaust but other sections of the waveform as well. Also keep in mind that lots of engines may have separate variable valve timing control for the intake and exhaust valve-train. So, the valve timing system being used is something to keep in mind.

Another analysis is that of points C and F. These two points should be the same height or pressure/vacuum. Points C is are the gases after the power/combustion cycle, but since the test is done without combustion, it's just compressed air. So points C is the original vacuum before the previous compression and power stroke and point F is the very beginning of vacuum creation, right after the valves have closed. So remember, these two points should match, otherwise there's leakage in the cylinder. The leakage could be due to valves, seats or piston rings.

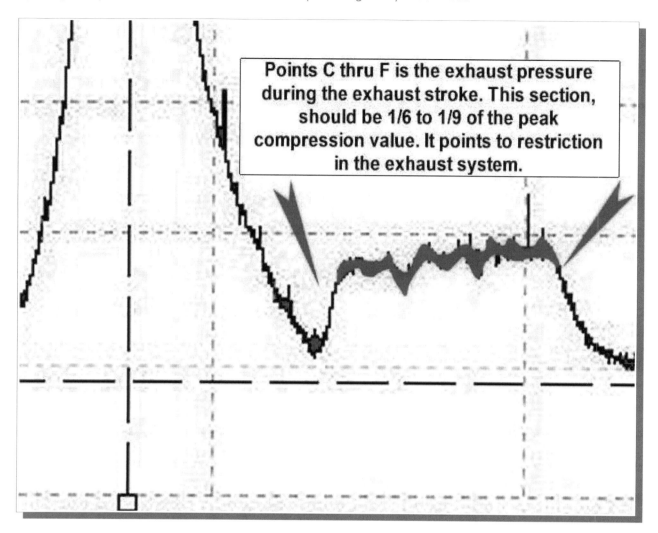

Finally, remember as mentioned throughout this book, that C thru F or the Ex line should be 1/6th to 1/9th of the peak compression or point A. Remember to always look for the relationship between points, not necessarily PSI levels.

Points F thru G with IN included is the intake hump:

Point F is the beginning of the intake stroke. Now point E is the exhaust valve closing and F is the intake valve fully open. The halfway between these two points E and F should correspond to the 360 degree TDC mark on the chart. The TDC mark can sway closer and further from points E and F, but stay within these confines. As with the previous points C and D, if the TDC mark is to the **LEFT** of point E then CAM timing is **RETARDED**. If the TDC mark is to the **RIGHT** of point F, then timing is over **ADVANCED**. Again, variable valve timing or a jumped timing belt/chain can cause issues like this.

Interpreting Compression Transducer Waveforms – (Including Comp-Peek-Prob

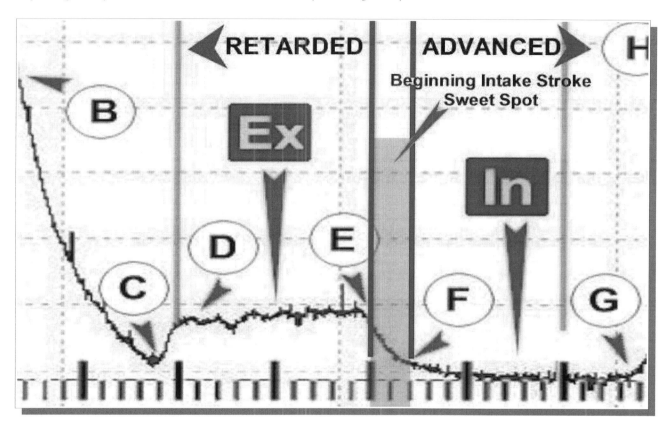

Point – G is the intake valve closing or fully closed. A few degrees before the BDC mark at 540 degrees the intake valve started to close. So from point G to the next Point A is the compression stroke. Afterward, everything starts to repeat again. The compression cycle work the same way as in the beginning of the analysis; it denotes the ability of the cylinder to seal and hold compression. It is an indicator of proper valve and ring sealing. Take Point G thru A and all the other points together and you'll have a great idea of what's happening in the engine.

Interpreting Compression Transducer Waveforms – (Including Comp-Peek-Prob

Included Comp-Peek-Probe Oscilloscope Software

Interpreting Compression Transducer Waveforms – (Including Comp-Peek-Prob

The Comp-Peek-Probe Software

Included with the Comp-Peek-Probe is the FREE Oscilloscope Software. This is a no nonsense software that allows you to view the engine compression waveform with the Comp-Peek-Probe. It is available together with the probe.

The newly developed Comp-Peek-Probe Oscilloscope signal conditioning hardware, with included free software is an industry first. We know most of you don't have the resources to invest thousands on an Oscilloscope just for when you use it occasionally. So, we went ahead and developed circuitry just for the Comp-Peek-Transducer-Probe. This oscilloscope **hardware** is only meant to be used with the Comp-Peek-Probe, not by itself. The Comp-Peek-Probe can be used with any oscilloscope without the included Scope software, but the scope part of it is meant only for the Comp-Peek-Probe. So you can't use the included oscilloscope by itself. It is a matched set, Comp-Peek-Probe-and-Scope.

Here we created the right electronics to work with the Comp-Peek-Probe. The Comp-Peek-Probe-Scope box and software runs on your Windows XP, 7, 8 laptop/desktop and has the right filtering circuitry and signal conditioning needed. The included free software gets the Comp-Peek-Probe signal and shows you the waveform, which you can save to the hard-drive if you want. The Comp-Peek-Probe Scope is a dual channel unit that you can synch to another signal, like a vacuum transducer. So, with this package you're set to diagnose and analyze automotive compression signals. All you need extra is a vacuum transducer if you want, but it's not needed.

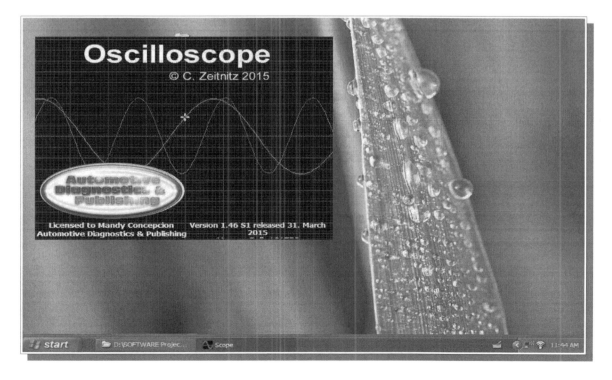

Interpreting Compression Transducer Waveforms – (Including Comp-Peek-Prob

Detailed Scope Software Explanation

The following section is an analysis manual for the Comp-Peek-Probe Oscilloscope Software. A detailed explanation is given here on it's features.

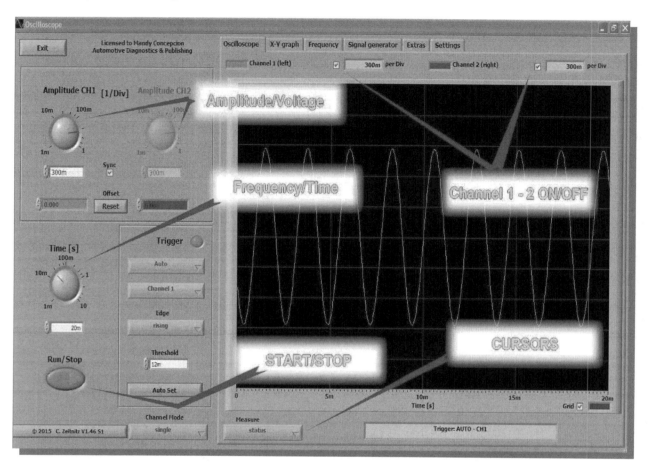

The Comp-Peek-Probe Software is a PC based Windows XP, 7, 8 & 10. The software runs on your sound card and the Comp-Peek-Probe connects to only channel one. You can use Channel 2 as a Synch channel, say to detect when the transmission shift-solenoid A is ON or when Coil-A fires, etc. The signal conditioning box protects and conditions your comp-Peek-Probe output line. It protects you PC from the outside world, like an ignition spark zapping the Comp-Peek-Probe signal wire.

Once connected, the Comp-Peek-Probe Scope software needs no drivers to install and is ready to go. The software also had other sections, such as frequency generator, but that's outside the Comp-Peek-Probe's usage and are implemented in our other equipment products gadgets, such as the PC Based Sensor Simulator.

Interpreting Compression Transducer Waveforms – (Including Comp-Peek-Prob

Signal Generator Section Output

Next are a few Snap Shots of the software and a brief short explanation of what the section does.

This is the Comp-Peek-Probe Scope Software with the signal generator output window. This section is not used on the Comp-Peek-Probe, but it is used on the Signal Simulator unit that we also offer, with the dedicated signal output circuit.

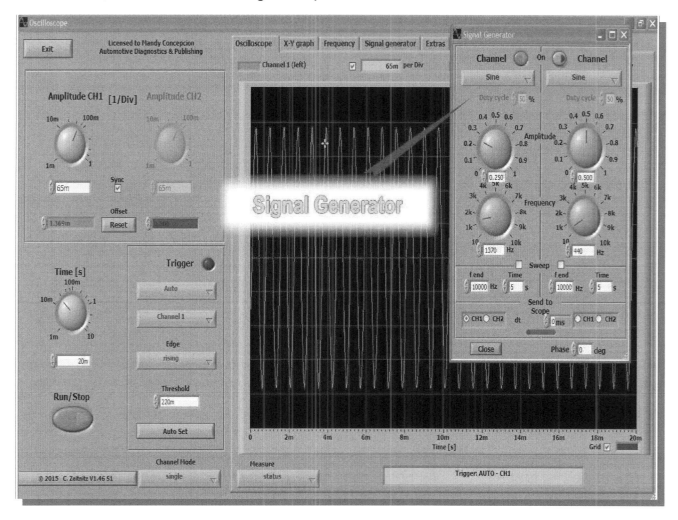

Such circuit is needed to adapt the output of the PC to your car and for protection. DO NOT USE the scope software by itself connected to an outside signal other than the Comp-Peek and Vacuum-Peek, or you may destroy your computer.

Amplitude Section

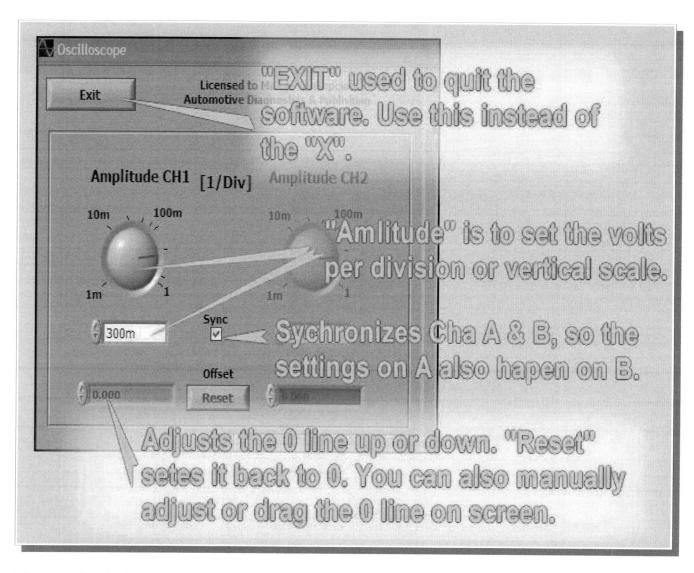

This section is the Amplitude or vertical scale. It is used to measure volts and is in volts per division. As we've said before, this is not necessarily the most important measure on the Comp-Peek-Probe's signal. What matters is the relationship between all the related measurement points.

Interpreting Compression Transducer Waveforms – (Including Comp-Peek-Prob

Base and Time Trigger Levels

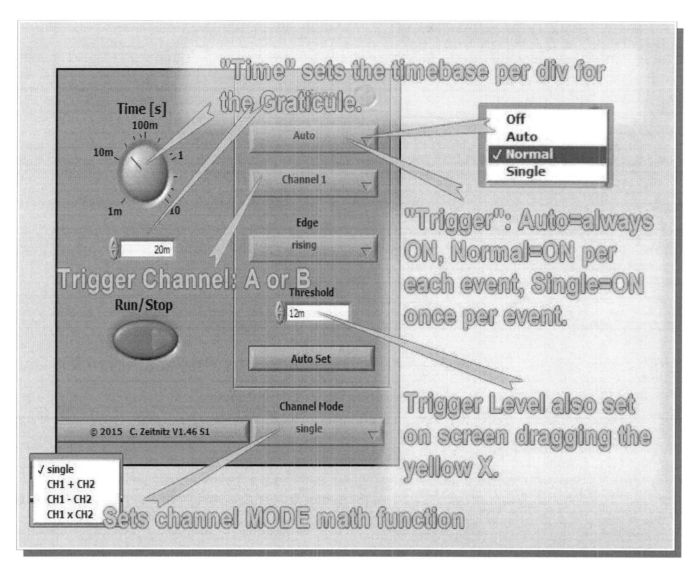

Following section is used to set the time based and the trigger levels. This includes the trigger channel for "single" shot to trigger ON the EVENT once, NORMAL or once for each event, and AUTO for continuous trigger. Say you want to trigger on each time the Comp-Peek-Probe detects a low level. Then you set the trigger level for the low signal level and set the trigger to single. Now each time the Comp-Peek-Probe reads a lower lever, the waveform will show up on screen. The possibilities are quite a few.

A Final Footnote about the Vac-Peek-Probe

A companion to the Comp-Peek-Transducer is the Vaccum-Peek-Transducer. They look more or less the same, but are not. The Comp-Peek-Probe is built to withstand high pressures on Gas and Diesel engines, while the Vac-Peek-Transducer probe is for minute Vacuum readings and can also be used with fluids, such as fuel the fuel rail pressure. Here's a brief explanation of the Vac-Peek-Transducer Probe.

Automotive Vacuum Peek Transducer Probe with included Oscilloscope Software

Measure dynamic vacuum and fuel pressures. Here's our super deal Automotive Vacuum Peek Transducer Probe with additional Scope Software. If you already have a scope then fine, but it can also be used with our Scope-1 and any other lab oscilloscope including Fluke, Vantage, Pico, Techtronix, etc. The Vac-Peek-Probe and Scope also comes with it's own manual CD-ROM video, which helps you understand how to use it. NOTE: The Vacuum-Peek-Transducer is not a gauge. It is a transducer and works by outputting a dynamic waveform; so the engine has to be running. It reads vacuum as well as fuel pressure changes. It also works at the exhaust, by reading the subtle exhaust back-pressure.

Why we created the Automotive Vacuum Transducer Probe (Vac-Peek-Probe)

There are a million and one situations where the Vacuum/Pressure Transducer is the only choice of tool to use or the one with the shortest path to a diagnosis.

Here's an example:

Your vehicle has a missing engine, sputtering, missing, shaky idle. You suspect it's a clogged injector after running some basic tests. So, pull out your trusted oscilloscope, Vantage, Fluke, Pico, or the Scope-1 and the Vacuum Transducer, with it's ability to sense fuel changes. Connect the Vacuum Transducer to the fuel pressure regulator vacuum hose or the fuel-rail itself. Start your engine and look for 4, 6 or 8 pulses or humps, as seen above. These corresponds to the fuel injector openings. If you see a clogged injector, then the humps are shorter or as seen above, no humps is visible. That's it... You found, without any reservations, that the issue is a clogged fuel injector. And remember, no dis-assembly required with the injectors in place.

Why you need the Vacuum Transducer

The Vacuum Transducer is a dynamic vacuum/pressure transducer that produces it's own signal. It is meant for "Fast Moving Changing Signals". It is not a pressure sensor for static pressures, like a fuel or vacuum gauge. It captures fast differential changes in vacuum or pressure. It is not meant to do compression tests; for that you use our "Compression Peek Pressure Transducer".

Interpreting Compression Transducer Waveforms – (Including Comp-Peek-Prob

The Vacuum Peek Transducer allows you to do injector flow tests, exhaust tests to detect burnt valve-train components, intake manifold vacuum tests to diagnose mechanical integrity, all kinds of small changes and many other scenarios where dis-assembly sets you back hours of work.

Solution to your problem

The Vacuum Transducer dynamic output (dynamic means fast moving) signal can me read by any scope/graphing meter, like the Vantage or any oscilloscope. If you use our company developed Scope-1, then you also get the "Signalator" software module. The "Signalator" is a software technology that reads and interprets the signal, and comes up with a solution. Other than that, the Vacuum Transducer comes with it's own video that lets you make some sense of what you're seeing, but experience is a must here. The Vacuum Transducer comes with all the needed cables and adapters. There's nothing else to buy, other than a different scope or graphing meter (it comes with it's own scope-software).

The benefits of the Vac-Peek Transducer

Using the Vacuum Transducer has one main benefit, it allows you to short cut hours from the diagnostic process. It is not meant for every occasion, but if it is, it makes for the best tool of choice. The Vacuum Transducer generates it's own signal. There are no connections to the battery or any other battery for that matter. The transducer's internal circuitry has filtering components to clean out the signal, in case you have a car with lots of EMF from the spark plug, wires or other ignition parts. The Vacuum Peek Transducer also comes fuel-fluid pressure ready. The BNC oscilloscope cable/connector and PC adpter is included, a BNC to Mic cable adapter is also included, as is a vacuum hose and a few other brass attachments. The unit comes with it's own protective plastic case. Feel free to right click and enlarge to images below for a clear view of the Vacuum Peek Transducer.

Following are photos of the Vac-Peek-Transducer probe.

Interpreting Compression Transducer Waveforms – (Including Comp-Peek-Prob

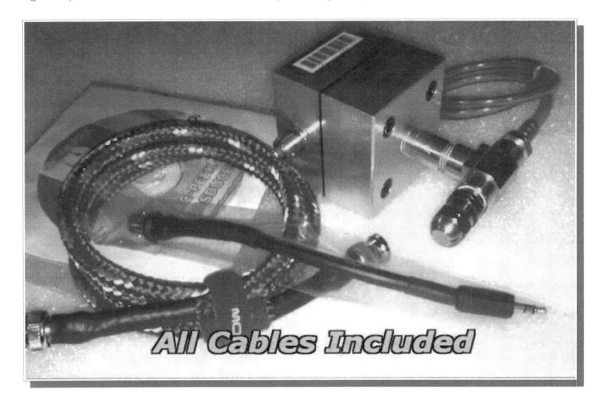

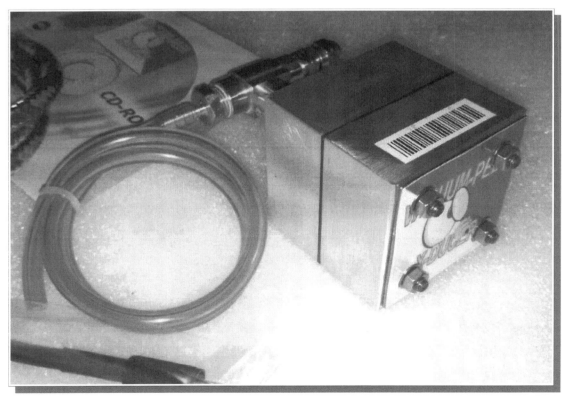

Interpreting Compression Transducer Waveforms – (Including Comp-Peek-Prob

Interpreting Compression Transducer Waveforms – (Including Comp-Peek-Prob

About Us and Mandy Concepcion

Integrating Technologies for the Shop of Tomorrow

Mandy Concepcion (Senior equipment and software developer, publisher and writer).

For the past 30 years, advances in automotive technology has NOT kept pace with the available training and technician integration/resources. The modern technician or repair shop is now faced with a great deal of decision making. Things like which equipment to buy, where to get trained, and how to reach the right repair information are everyday dilemmas that have to get solved for businesses to be profitable. Techs and repair shops throughout the country are always scrambling to make their job easier. The concept of integration is almost unknown in auto repair circles due to the fact that many companies keep their technology proprietary. Integration is the ability of all shop applications/programs to talk to each other. This means that your invoicing system should know from the get-go that if you're faced with an O2 sensor DTC, the scan tool talks to the invoicing system and it's able to detect if the same O2 sensor was replaced a month ago, without the service writer having to check manually at the historical records. The previous example demonstrate one of thousands of possible scenarios dealing with the concept of integration. In order to tackle the super complex vehicles coming out today and tomorrow the modern repair shop needs all the help they can get. Integration between the scanner, techline-service, invoicing, oscilloscope (measuring equipment) and the Internet is an absolute must. Integration is meant to lift a huge load from the technician and service writer thereby increasing efficiency.

There is also one area in which this industry has always been very lacking, and this is the ability for a tech or shop to get the right training and equipment. The right training doesn't mean a text book and useless Kits that simply tech you what you already know. The modern auto technician needs specific and specialized training. In other words, what do you, the tech or repair shop, need to know to fix a modern vehicle system. The same goes for the equipment. As time goes by the equipment and information on how to use this equipment is not keeping up with time. Companies increase and set prices and capabilities as they see fit, with no regards for what the technician needs.

AUTOMOTIVE DIAGNOSTICS & PUBLISHING

Here, at AUTOMOTIVE DIAGNOSTICS & PUBLISHING we do just that, offer modern, practical, and up to date training for the automotive repair community, as well as the right type of equipment and how to use it. We are very much interested in allowing the modern repair shop technician squeeze as much from the diagnostic tool as possible. We are working very hard to always implement an entire array of HELP features into our PC based tools. Such features will comprise embedded videos right into the software, more information, automated testing of different components, parameter red flagging for off-scale values, on-the fly diagrams for different components and wireless communications for easy data transfer. Integration is also a must for the shop of tomorrow. The ability of our Scan-1 or Scope-1 to talk the invoicing system and to link to the Techline-Service over the Internet is an essential part of our technology. We cater to the modern repair shop and/or technician with the best in PRACTICAL TRAINING and PC Based EQUIPMENT through our books and DVDs, software and interfaces. Our books and videos are custom made/edited and geared towards the practical, hands-on aspect of automotive repair. You get top performance training in the comfort of your own home, for a fraction of the cost of personalized classes. And through the VIDEO DVD and CD-ROM media, the possibilities are limitless. All our videos are packed with special effects-titles and explanations making then a whole lot easier to comprehend. At the same time, we also have Training capabilities above and beyond what any other company offers. Through our training the technician can link with our instructors and receive personalized or group training using the Internet or one-on-one. This is one level of integration that we've been able to achieve, which is second to none.

Don't be caught off-guard in the automotive repair training, information and equipment battle. Trust our experience and our products to supply you with the best SPECIALIZED service and training possible. Thank you.

Who is Automotive Diagnostics & Publishing

In 2003 Automotive Diagnostics and Publishing was founded by Mandy Concepcion. The goal was to provide integration, custom geared training, repair software and equipment for the automotive repair community. The pace of modern automotive technology is almost incomprehensible to most people outside the industry. Few professions are so demanding in both the amount of training and equipment needed for the job. Our company understands that and engages primarily in the development of features and information rich content training, repair and equipment products.

Integration, which is now the center of many of our efforts is the ability for smart applications/programs to talk to each other and come up with solutions that would otherwise be impossible to detect.

Imagine if your shop had all the basic technologies working together; the scan-tool, measuring equipment (scope/DMM), invoicing system and Internet techline service. Whatever happens on one application the others will also know and recognize a possible issue or problem. This is what you can expect from our technologies.

As a wholesale operation to schools and tool houses, we develop, test and manufacture all of our dedicated PC based software, interface boxes and training DVDs, books and CD-ROMs. We do not sell third party products, although that may always be a possibility. Since our company is also deeply immersed in actual diagnostics, our products are made to work. We know what the repair shops need and we keep it in mind throughout the design process.

Innovative Introductions

The introduction of the Scan-1, Scope-1 and Virtual-Training Engine built into our software has given us an edge in technology. These are the most modern and feature rich products on the market, with an eye to a short learning curve and ease of use. On the training side of our operation, our DVD videos are constantly being updated and new ones are being developed and coming on line all the time. No other company can match us in expertise, knowledge, complete contact with the industry's need and quick manufacture. Due to this fact, we cater to the specific needs of technicians, shops owners, schools and companies like no other tool developer. Although we cater to the average technician and repair shop, schools, specialized diagnostic entities and government institutions in custom design of specialized products.

We also run and operate a Nationwide Web-Based Techline Service. Our Integrated Scan-1 and Internet Techline Service are technologies created at our California facility. Our job is to help repair shops solve faults that are difficult to solve. This way we can stay current with all the newer systems and recurrent problems out in the field. This is by far the biggest difference between us and our competition, we are engineers and working technicians. The company also engages in training on-demand, tailored to the specific needs of each shop using the Scan-1 and Techline Internet service. Our on-demand training programs can range from equipment usage (scanners, oscilloscopes, GMM, signal generators as well as our own PC Based equipment), OBD-2 strategies, tracing network faults, factory antitheft, ABS systems, Electronic Transmissions, to how to advertise your repair shop on the internet. The repair business is changing at an impressive rate. Future shop owners will have to be on top of technology trends, such as attracting customers using a Website, using the PC to do their diagnostics and having the right information and training. Technology integration and smart application for your shop will give you an edge and make your customers appreciate your business.

Interpreting Compression Transducer Waveforms – (Including Comp-Peek-Prob

Made in the USA
Middletown, DE
08 February 2021